HERE WE GO ROUND
THE MULTIVERSE…

WENDY SHUTLER

Hamilton Brown & Skelding
Cornwall

First published in Great Britain 2002 by
Hamilton Brown & Skelding

Copyright © Wendy Shutler 2002

The moral right of the author has been asserted

All rights reserved. No part of this publication may be reproduced, stored in a retrieval system, or transmitted in any form or by any means, without the prior permission in writing of the publisher, not to be otherwise circulated in any form of binding or cover other than that in which it is published and without a similar condition, including this condition, being imposed on the subsequent purchaser.

A CIP catalogue record for this book
is available from the British Library

ISBN: 0-9542717-0-X

Edited and Produced by Phoebe Phillips
Printed and bound by MRM Graphics

Hamilton Brown & Skelding
Whitehay, Withiel
Cornwall PL30 5NQ

To the memory of my cousin
ROBIN BENNETT

SILVER SCRIBBLES

I'm trying to identify
what slips away to hide, to try
to understand and simplify,
extract its essence, hope to find
luminous, defining lines
to scratch the surface of your mind,
as vapour trails from jets up high
leave silver scribbles on the sky...

Acknowledgements

First and foremost, I must thank Roger Hamilton Brown and Simon Skelding for taking the bold and generous step of publishing this book, and for doing so with such enthusiasm and style. Also Dick Douglas-Boyd, whose invaluable help has been indispensable, and whose opinion matters more to me than he probably knows. Many thanks to Derek Robinson for writing the foreword, to Tom Robb for his beautiful cover design, and to Judy Douglas-Boyd for her tremendous unfailingly cheerful support, not least for organising such delightful company meetings.

Very special thanks to my editor Phoebe Phillips who has not only been a guiding light, but has done for this collection what a top hair stylist should do for a head of healthy but slightly unkempt hair. Every suggestion she has made, each precise snip of the scissors, has been for the better. And for her guidance in taking the manuscript through production to the finished copy.

I am extremely grateful to Alan Bennett and to Michael Frayn for their wonderfully encouraging remarks, and their generosity in allowing me to quote them, and to Michael Blakemore, for his interest, and for bringing "Quanta" to Michael Frayn's attention in the first place.

Last but not least, I should like to thank those many friends who have shown such interest in these poems from the start, in particular Madeline Smith, April Turner and Laura Turner.

Wendy Shutler

Foreword

Lawyers have a saying "the devil is in the detail", meaning that a big case can turn on something in the small print. It applies to the arts, too. It certainly applies to fiction. Every novel I've written has one sentence - maybe just one phrase - which, to me, is worth all the effort of writing the whole damn book. Of course, there's no way of making that handful of words without producing the entire novel. The devil is in the detail, but the detail can't exist alone.

And that's very true of Wendy Shutler's poems. They are rich in devilish detail. What's more, I know of no other poet who is so at home with scientific ideas. My knowledge of science is pitiful, but for me she flings open a window when she finishes her long and funny poem about quantum physics - which we all experience when we see a blank yet live TV screen covered by a snowstorm of dots - with a reminder that this is light that was created at the very birth of the universe and is still radiating everywhere;

> Cosmic background microwaves
> float through us, eternal rays.

Those are eight words that span all time since the Big Bang. Nice work. And she is wonderfully economical about places - spiderweb span of the Severn Bridge - and about people:

> When we bit the apple and we found
> that we weren't always going to be around...

Some things can be best said in poetry. Wendy Shutler is very good at saying them. I could quote many more, but what's the point when the real thing is waiting to be read? Turn the page, and enjoy.

Derek Robinson

Introduction

Born in South Gloucestershire, where generations of my mother's family, the Beazers, have lived since before the 17th century, I grew up in a suburb of Bristol, with my parents, grandmother, brother, a dog and two cats, an extended family of aunts, uncles and cousins close by. My parents were gregarious, sociable people. A tumbledown holiday bungalow overlooking the Bristol Channel was shared by everyone. Many years later, I bought a caravan near the old bungalow for a weekend retreat. Although I love London, it's wonderful to have a home in the country too, especially in summer. I have a little garden, and a view over the sea to the Welsh hills - a marvellous, quiet place for writing and watching the changing light on the water.

My mother, her life as an artist cut short by the convention of the day that married women did not work, encouraged me to choose a career. To learn to "speak nicely" I was sent to Mrs Phyllis Hill's when I was 12 for elocution and drama classes, and to iron out my Bristol accent. There I discovered acting, that heady, anti-shyness drug, an addiction which eventually led to my first theatrical contract as a student Assistant Stage Manager at the Little Theatre, Bristol.

The exams set by the London Academy of Music and Dramatic Art included poetry speaking and its technique. But Mrs Hill taught more than iambic pentameter and voice production. She inspired her pupils to appreciate the beauty and resonance of words. We saw every production of the Bristol Old Vic at the Theatre Royal, and performed short plays once a year at the Parish Hall - I was in them all! It channelled the passions of adolescence into a head full of Shakespeare, poetry, and the theatre.

Poetry expresses those things common to us all which are not often talked about in ordinary conversation. Just as acting enabled me, disguised as a fictional character, to do things I would never dare, writing poems enables me to say what I believe to be true, and communicate those beliefs to others. We all lie, at least by omission in everyday life, if only to avoid confrontation. But poetry can touch emotional depths, make us aware of our common ground, and lessen feelings of isolation.

I don't remember when I began writing; if my head's in a muddle about anything, from the pangs of love to the complexities and paradoxes of quantum physics, writing everything down and then sorting it into some kind of order seems to clarify it in my own head. Not naturally neat or organized, I'm untidy and always late, yet for most of my life I've been an actress, which requires discipline and precision, and often a stage manager, which depends on organization. Like all actors, I've done a whole raft of other jobs - barmaid, model, shop assistant - and more useful, perhaps, writing code-frames for market research analysis, which organizes a pattern out of the comments of people interviewed for surveys. Perhaps this too has helped me to sort out the patterns in my own head formed from a wide variety of interests.

Acting came first, probably the most important area of my life, and yet it's strange that I've never written about the theatre. It's too much like poetry itself - holding a mirror on life - and you can't go on holding up mirrors of mirrors ad infinitum without blurring the edges. No doubt I was ready to look through other mirrors when my cousin Robin came home for Christmas one year, and could hardly look up from reading "In Search of Schrödinger's Cat". He lent me his copy when he went home, and the world has never looked the same since.

Once I delved into the story of quantum physics, I was hooked. We take for granted computers, lasers, even nuclear weapons, and yet these things exist because physicists understand something (although never all!) about the way the universe works at the level of fundamental particles. I learned quickly that subatomic particles operate in a way that defies common sense, and raises astonishing questions about ourselves and our world. There is more magic here than in any science-fiction or fairy tale.

Read John Gribbin's fascinating books, "In Search of Schrödinger's Cat" and its sequel "Schrödinger's Kittens", both written for the non-scientist. Michael Frayn's extraordinary play "Copenhagen" will stretch the mind to accommodate a new, incredible world. The poems in the first section of this book were written to help myself to understand the implications of quantum physics which among other things have led to the technology we so take for granted, including the word-processor I'm using at the moment, and I hope they also convey a little of the excitement and wonder I found. That stars are light from aeons ago, from a time before life appeared on earth is surely wonderful, and they still sparkle!

Few things make us sparkle as much as that microcosm of life, romantic love. From the torture and joy of its birth to the devastating heartbreak of its death, it's beyond reason. But for me it helps to tame the beast a little by writing, which at the same time lets everyone know they are not alone in their madness. The wider world, too, the places I have visited, the sudden catch of breath at a moment of sorrow or beauty in the countryside or the city - these have all been inspirations from this particular branch of the multiverse in which you and I live.

My cousin Robin would have loved to have seen this collection of poems, but he died in 1999, and so I dedicate it to his memory.

Wendy Shutler

CONTENTS

Acknowledgements
Foreword: Derek Robinson
Introduction

PART ONE: MULTIVERSES   10
Quantum Questions
Searchlight

PART TWO: UNCERTAIN VERSES   32
Lovelocked
Thoughts
Places

PART THREE: AND THEN...   68
Buried Treasure
Red Shift

A word about words:
a glossary of scientific terms   90

Index of titles   94

PART ONE: MULTIVERSES

QUANTUM QUESTIONS

SEARCHLIGHT

QUANTUM QUESTIONS

THE MULTIVERSE

Each moment is a flash of light,
a rainbow-prism glancing
from one facet of one crystal
out of many dancing,
glittering in the chandelier
of our universe,
hanging among many more
in the multiverse
of worlds whose waves can never break
upon our shores, within whose wake
we live, yet cannot know,
from whom we too are hidden.

Each trapped in our own world and time,
we travel paths unbidden
through hyperspace, reflecting in
its endless mirrored halls.

We only know one flash of light
and think it tells us all.

DAYBREAK

Sunlight breaks upon the sharp
and jagged edges of the dark
and broken mirrors of the world;
splinters into colours, swirls
into eyes and into brain -
imprinting there a memory-stain -
then out across the dark in waves;
photon-angels.

Our world is but a narrow strand,
a seven-colour rainbow band
where runs the line that we call time,
along whose tracks we are confined.
Blinkered little creatures, we,
who only here and now can see,
yet sometimes seem to see (briefly)
beyond the rainbow.

QUANTA

Everything exists, is there -
that polished table, this old chair -
regardless of whether you're looking or not.
Now is now, no matter what;
yesterday won't reappear
and tomorrow's not yet here.
That's the world, according to
The Local Realistic View.

When something moves from here to there
it surely must have gone somewhere?
Measuring electron spin
will prove that common sense must win.
We can trust reality
if the answer's Bell's Inequality.
There has to be locality -
otherwise, where would we be?

But the calculations show
this is actually not so;
when crucial measurements are taken
common sense is sorely shaken.
Bell's Inequality is breached
by any photon out of reach.
It's really neither here nor there,
or maybe it is everywhere.

For electrons will behave
as particles and then as waves.
Wave-particle duality
means here and now is all we see;
waves of darkness crystallise
in the headlamp-light of eyes
and melt away. It isn't true,
the Local Realistic view.

Nothing exists, there is nothing there
if no one is looking, no table, no chair.
For you can only say for sure
where they were when you last saw -
in the past. And where is that?

Did anyone ever find Schrödinger's Cat?
If yesterday's somewhere in space,
tomorrow is another place.

Our perceptions only show
what it is we need to know.
We're equipped with what we need
in order to survive and breed.
What worlds may pass while we take snaps.
freeze-framing waves as they collapse!
Could we but stop the clock and see
the ripples of Uncertainty!

Probability and chance
interference patterns dance
upon a screen; they come from where?
Can one little dot be everywhere?
Parallel worlds, the future, the past,
all here together, forever to last?
Or when you're not looking it all goes away
and nothing is real, in any way?

Our ship cuts through an ocean sea
of waves of probability.
One little dot of each we see
of every possibility,
while rippling, spreading out of sight,
all that was, will be, or might
have been, beyond our blinkered view
where hopes and fears and dreams come true.

And if you think this can't be true,
think again, next time that you
watch "Eastenders" on TV
in colour, buy a pack of tea
price checked by a laser-scanner
in usual supermarket manner,
for these things could never be
but for quantum theory.

It's very likely that the chair
and the table will be there
when you turn to look again,
and they make pictures in your brain -
for their clarity selected
interference waves rejected -
but this censored TV show
is all that you can really know.

Some say one day it will all contract
to a singularity, re-enact
the Big Bang, but in time-reverse
of this expanding universe.
(But if there **is** no absolute time
is it happening now? For time
is never "wrong" and never "right",
but relative to the speed of light.)

From the light of that first dawn
when the universe was born
radiation is still found
uniformly all around.
Cosmic background microwaves
float through us, eternal rays,
rippling through, sometime, somewhere,
and one little dot **is** everywhere.

MANY WORLDS

All those ghostly, wavy, worlds,
liquid like the sea,
this one I chose to look upon,
this is real for me.
I make a pathway through the waves
of probability.
Where I touch, they crystallise,
become reality.
The Red Sea waves I parted
to make a space to be;
I cannot return, and they
can never follow me,
those other worlds where Jonny runs
and skips, and he can see,
where I was never born, and where
you died so tragically.

All those ghostly, wavy worlds,
liquid like the sea;
everything that might have been
just beyond may be
dancing in the shadows,
each possibility.
This one I chose to look upon,
this is real for me -
where Jonny in his new wheelchair
hears music on CD,
where I write words you read or hear
wherever you may be
and from the ripples just ahead,
jostling in the sea,
I must choose tomorrow,
irrevocably.

VIRTUAL REALITY

On the scale of the very vast,
where speeds are incredibly fast,
there's no such thing, says Einstein,
as an absolute moment in time.

On the scale of the minuscule,
where quantum uncertainty rules,
Heisenberg says there's no such place
as a definite point in space.

There is no time, there is no place;
we're mostly made of empty space.
Very far or very near,
it seems we are not really here.

LEAP BEFORE YOU LOOK

Purposeless, we blindly leap
from here to anywhere
by way of nothing, through nowhere,
via everywhere.

On unknown future islands land
across transparent seas,
invisible, infinity
of probabilities.

We cannot know the way we go
for we only see
sharply, interference-free
where we used to be.

So little, micro-time ago,
could we not call it "now"?
How many dots of space-time go
to make the instant "now"?

Is time itself a series
of particles observed?
A subatomic spacetime curve,
digital, unblurred?

Each defining moment's all;
observation's what defines
the fleeting instant, present time;
no grand design.

If here and now is all there is,
that photon passing through both slits
 is outside time and can't exist
in this hypothesis.

Moments tangled, moments stretched
on a web of electromagnetic thread
where long-dead stars are shining yet
overhead.

How long is now? How deep the sea
whose waves of probability
are frozen by a glance of eyes
and for that moment, crystallised?

When, like Uncertainty itself,
I grasp at these ideas,
they slip beyond my reach, I feel,
and worlds conceal.

EVERYWHERE AT ONCE

You accept, when you hold a phone in your hand
you can speak to a friend in a far-off land,
resurrect the dead, by video,
Rudolph Valentino or Marilyn Monroe,
store information from all over the place
yet take up hardly any space.
So why don't you believe that a thing could be
everywhere at once, when you don't look and see?

Shadows ripple out of sight,
freeze beneath the glancing light
of observation; crystal clear,
particles, not waves, appear.
The God Observer looks away.
Do they resume their dance, to play
in eternity? or die,
outside the spotlight of God's eye?

This is what electrons do
in every atom of which you
and I and everything are made.
You with the laptop and the microwave,
your digital, state of the art TV,
and made-in Japan technology -
if the atom bomb had dropped on you,
would you still doubt this could be true?

FOOTPRINTS IN THE SNOW
(thoughts on Transaction Theory)

Waves that travel faster than light
catch up with the past beyond our sight,
return from tomorrow to show us, today,
the path they have beaten for us - the way.
The road we must take already decided,
we follow the tracks that time has provided.
Tomorrow, making safe today
with the help of yesterday,
creates each present moment, guides
where past and future meet, collide,
counting to zero, the way to go
following footprints in the snow.

STAY TUNED

If wave by wave we choose our world
from many, others might
exist at other frequencies
beyond the speed of light.
The speed of light is constant. Well,
imagine otherwise;
if photons bounced from people's lips
faster than from eyes.

As he moves towards you, first
the disembodied lips,
appearing in the air, approach,
a flying, floating kiss;
and then the eyes, and bit by bit
his face would be assembled -
only to disintegrate
with his slightest tremble.

Everytime you to try to take
a walk along the street,
a broken-jigsaw blizzard swirls
from everything you meet;
blinds you with a tumbling blur
of interfering waves -
could evolution have occurred
if this had been the case?

If we see pictures frame by frame
like on a TV screen,
blind to the confusion
swirling in between,
why not evolve to only see
photons at certain speed,
forbidding any other,
according to our need?

Passing on genetically
a narrow, tunnel vision;
inheriting a tendency
to channel with precision
a pathway through the clouds of Other
we choose not to see;
selecting definity out of infinity,
time from eternity.

MAGIC CIRCLES

No particle can send a shiver through
your spiderweb. You send no e-mail now,
nor pick up what I send to you. Or do you?
Your website on the Internet's shut down.
Electrons of your clever brain are still,
no longer jump for their allotted share,
their quantum of energy to spark and to fill
like exotic goldfish in a jar,
darting up to take a breath of air.
Is a wave a spirit of some kind?
Are particles the world that we know here?
Electromagnetism knows no time.
In magic circles, waves of long ago
are out there still, and part of you, I know.

SEARCHLIGHT

SEARCHLIGHT

Why choose darkness over light
of knowledge, raising yet more new
undreamed-of questions?

Do you see no mystery, no romance,
no wondering what it is all about,
in science?

Does every probability
exist, or only what we see?
There's mystery!

We and our fellow-creatures
are made of the stuff of distant stars.
That's romance.

Surely it's better to search in the light
than to grope about in the dark at night
to find something.

PANTHROPIC PRINCIPLE

The universe is the age it is,
and at the size and stage it is,
because if it were otherwise
we would not be here, have eyes
to see, or any way to know
that the universe is so.
Does this put man at the centre of things,
make homo sapiens sapiens king?
No! It's so for one and all,
for all creatures great and small,
each with their own point of view,
equally valid, equally true.

The blade of grass to synthesize
light reflecting to our eyes
the green hill, and the woodland glade,
the old church tower, or crumbling cave
that echo to the bat, who sings
silently at dusk, and wings
his way among the echoes from
his own vibrating, silent song.
Each species finding different ways
through the misty, murky maze.
None of us would live today
if things were any other way.

BEYOND OUR BRANE

We die every moment;
the moment I wrote "we" has passed
already. And the me
that I was then is dead and gone.
Or ripples, races out and on
forever, as a wave
who waits, perhaps, to be recalled,
squashed into a mould
by someone's eyes or someone's ears.
But who is there to see or hear?

Vanished moments, prayers and dreams,
all neatly stacked in layers, it seems,
like annual rings of trees.
Where are we? And who can see
at once, this solid tree
where layers of time expand, to follow
yesterday out of tomorrow?

What strange colours shimmer?
What unimagined sounds vibrate
and swirl among those layers
that peel away, that we may see
each moment, and a memory?
As this moment flies
though I must let it go, I know
at least that, thanks to you,
this poem is no longer dead,
but living now inside your head.

DIAMONDS IN THE SKY

In the cold hard glitter of a frosty winter night,
they stud the sky like diamonds, the stars,
piercing time and darkness from afar.
Little star, I wonder what you are…
children sing.

Some are long-gone, burned-out, dead by now,
ashes to ashes, dust to carbon dust.
Scattered to the solar winds, to gust
to distant planets, buried, not to rust,
but fertilising.

Carbon, stardust, seed of life on earth,
star-children growing in the planet's womb,
we are hybrids, grown in catacombs.
A star's tomb, a diamond's mother's womb -
same thing.

The purest form of carbon is the diamond,
fruit of stars, cut and polished, bright,
flashing blue-white ice-light,
as if a star were stolen from the night
to set in a ring.

EGO TRIP THROUGH A BLACK HOLE

The galaxy that curves around me
shrinks to one bright star.
Enveloped in the dark,
I'm crushed into a quark
then to nothing.

But black holes eat up space-time,
turn it inside out.
A video rewinding,
in Dark of Darkness hiding,
to the start.

Running backwards, getting younger,
being unborn,
here where none can be detected,
here the dead are resurrected,
to live again.

Do they know they're going backwards?
Do we know we're not?
Tomorrow may be yesterday,
maybe we are here to stay -
trapped.

Either we're all everywhere
at once in every time,
or only for a moment live,
this instant all there really is;
that's it.

But if black holes are everywhere
in every galaxy,
running copies of us all,
then who is the original?
It must be me.

NOTHING

I am cold, black endless space;
for me there is no time, no place.
Could I be aware, I'd know
that somewhere out in the dark and silence
there, for an instant, glows
a globe of seven colours
dancing circles round a star
upon whose surface for a moment
living beings are,
who see and feel and hear
such things that I could never dream.
I would envy, if I could,
each sparkle in that stream.
What must it be, to be?
But I am cold, black endless space;
for me there is no time, no place.

THE MAGIC GENE

When we bit the apple, and we found
that we weren't always going to be around,
but only here for a limited season.
not for any particular reason-
life without hope;
we couldn't cope.

We told of spirits in the forest glade,
gods, and devils, nymphs and dragons made,
telling tales around the fire
myth and magic to acquire;
and those who believed
survived to breed.

In our own image we made God, that He
take from us all responsibility,
and send His son to earth to save
ourselves from earth of our own graves.
A little white lie
to help us get by

long enough to raise a generation
confident of ultimate salvation.
Until we wonder if it's true,
begin to question long-held views,
and wonder why
we all must die.

Rather we should wonder that we live,
we lucky ones, who evolution gives,
against the odds, this chance to be.
Why ask for all eternity?
This fruit is sweet.
Take, eat.

DIVIDED WE STAY

Science and spiritual matters may
pursue one truth, my friend, but they
are separate, like black and white,
tightly focused, sharp and bright.
Side by side, divorced, appear
patterns, well-defined and clear
as shadows in the southern sun -
but please don't let the colours run!
For black and white then turn to grey -
a fog in which to lose our way.

WHITE HORSES

Tomorrow-waves are rolling in,
bigger and bigger and bigger until
today's white horses break and curl,
crashing beneath today,
sliding to yesterday,
merging with distant tomorrows
rolling in.

PART TWO: UNCERTAIN VERSES

LOVELOCKED

THOUGHTS & PLACES

LOVELOCKED

FIRST LIGHT

Suddenly you struck a spark
and I could see through all the dark
beyond the farthest hill.

Occasionally, we aspire
to know that timeless, clear white fire
encapsulating all, to be
one with all Eternity,
the Light of Light, who can bestow
the secret that the flowers know.

Perhaps the light of that first dawn
when the Universe was born
burns within us still.

AMBER LIGHT

The sweet shock I feel when you enter the room,
the sound of your voice vibrates in my womb.
When I dare to meet your eyes,
all else vanishes and dies.

The flame that flared when, suddenly,
you looked at me so longingly,
kindled fires in me that day
that go on burning me away.

I'm afraid to let you see
the extraordinary effect you have on me;
and I see you fear me too.
I wonder if it's the same for you?

Drowning in each other's eyes
until we're lost and hypnotized -
this much we dare, yet cannot risk
a brush of hands, the lightest kiss.

Afraid to touch, connect, reveal
honestly the way we feel,
break barriers that barely hold
in check a powerful flood. Be bold!

Lips, hands, tongues; caress, explore,
lead us further, deeper, more,
release the rising, flooding tide.
Don't just fuck me with your eyes!

JUST LOOKING

Your sea-grey eyes pour oceans into mine,
and they make me giddy, like too much wine.
I'm drunk with you, you go straight to my head.
Will I sober up, if we go to bed?

PALPABLE HIT

Your eyes were hard, the pupils narrow,
shooting passion-poisoned arrows,
darting, straight on target, sure,
to my unsuspecting core.
The wound spreads slowly, deep in me,
poisoning me fatally
with love.

OUT OF ORDER

I never wanted, looked for this
aching, longing restlessness.
Where does it come from, this confusion?
Is it madness, some illusion?
Don't tell me this isn't real,
this new love I so strongly feel.

This is what's alive and real,
indestructible as steel,
powerful as a laser beam
cutting through the clouded dream
of ordered lives of day to day,
separately boxed away.

This love is all that I can feel,
so don't tell me it isn't real!

RED LIGHT

I'll walk in Regent's Park
before it gets too dark.
I need fresh air and a change of scene
to think straight.

Autumn trees glow;
in veins of leaves flow
pulses of fire, electric, intense,
red as blood.

Burnished copper tints
recall the sheen that glints
in your hair. The amber air
begins to blush.

Crimson, drenched with dew
at the thought of you,
bloom the dusky, velvet, dark
late roses.

Everything I see
keeps reminding me.
And I thought I'd see by a clearer light
after last night!

TRANSFUSION

Hold me. I cannot cope.
Hold me. There is no hope.
Part of me is dying, gone.
You brim with life and health. I long
to feel your heart beat into me,
pulsing, throbbing, feeding me
with love and lust sustaining.
I know you're not complaining -
you course through me as if we were
like Siamese twins, together;
I the weaker of us two,
drawing life and blood from you,
my life-support, so virile, strong -
I won't need you this much for long.

And when I've bled you dry,
I'll leave you wondering why
I had to say goodbye.

JUST SEX

Delicious agony so much
I'm rendered helpless by your touch;
I dissolve, disintegrate
in sweet sharp waves that break and break.
What joy I see it brings to you!
Your helplessness at my touch too.
I cling to you to keep me here,
and yet you make me disappear,
flying out, beyond, for ever,
nowhere, everywhere, yet never
closer to the earth, at one,
surrendered, naked, lost and won.

RICE PAPER MOON

Sharp and sweet like stolen apples,
secret afternoons
in darkened rooms

under the ghost of the daylight moon;
it looks like a fragile thing,
rice-paper thin,

a peeled-off disc of moon-skin
from the solid bright
moon of night,

luminous with borrowed light.
But rounded, real, it gleams,
however it seems.

Sustaining, ripe, no flimsy dream,
an apple thrown too high
caught in the sky;

really just a satellite
trapped in orbit, round
and round and round,

never ever touching ground.
Aphrodite's apple,
sunlight-dappled.

Sharp and sweet like stolen apples,
secret afternoons
in darkened rooms.

GREEN LIGHT

Your eyes are not the green of jade, exotic,
but of the light of a woodland glade
on a June evening.

Pale green light shines into me. Magic.
And I light up inside, and see
things differently.

We do not know we cannot stay, don't realise
now is the longest, lightest day,
the solstice.

Caught in a summer forest spell, enchanted,
in our leafy dappled dell,
treading dead leaves.

For you will turn away for good, you'll leave me
to find my way through darker woods
alone; set free.

THE SENSIBLE THING

I was unhappy, and you were kind.
That's why you're always on my mind.
You're right.
Doing the sensible thing.
Who cares if you don't ring?
We're just friends.

Why should you obsess me so?
There's nowhere for us to go.
Nowhere.
Even if you were free
you would be wrong for me;
I don't want you.

Anyway, I've got someone, too.
He's far more suitable than you.
What the hell!
But I wish you'd get in touch,
and I miss you very much.
Don't know why.

CONTROL FREAK

I've escaped from love's distractions
to a different world
which has plenty of attractions
I enjoy without
being side-tracked by some man
I love too much.
No more forsaking every plan
I've made for me!

I have an independent life
I do not sacrifice
myself to be a loving wife
or mistress; I am free.
I can do without that dream.
Smugly, so I thought -
then love struck like a laser beam,
precisely.

"Do not give him another thought"
is my friend's advice.
"If you do, it's your own fault!"
Yes, I know, I know!
Does she not have the slightest inkling,
could she have forgotten
that it has nothing to do with thinking
or whether he's worth it or not?

As if a thought of him would feed
the fever, make it grow
into a deep despairing need,
a physical addiction;
as if he were some lethal drug
that she would never take.
She sounds so clever and so smug,
like me, before.

BLACK WIDOW

In my mind, you were the cream,
rising to the top of the millk;
if I shake this bottle, I thought,
perhaps you'll homogenise
and I'll chase you from my dreams.
But you won't go away.
If you knew, you'd be afraid,
and rightly so.

Beware of me. I set you snares,
spin devious webs to entangle you.
I want to make you part of me,
to fill my emptiness with you.
Beware of me. Beware
of my fierce female passion.
I'll engulf you, devour you, show you that I care -
and call it love.

SEPARATION

Suddenly you'd gone,
torn from me in a great scream.
Then I was numb and in a dream;
normal rhythms seemed to stop;
time went by without me; shock,
nature's anaesthetic, blessed
me with a spell of nothingness,
and suddenly, I'd gone.

I was paralysed.
A shallow stream that babbles, flows,
blocked with pebbles, over stones
to sparkle on the surface, chatter;
nothing any more will matter.
The stones and pebbles shift again
like jagged broken glass, the pain
insisting I'm alive.

DEPRESSION

Drowsy luxury
of sun on skin
as pale translucent
sea shimmers.
Warm, kind
happy faces
as teacups clatter
in fragrant garden.
Yet inside,
the sharp ache,
the silent scream.

ESCAPE

Recaptured by the morning light
breaking sweet relief of night's
escape in Technicolor dreams -
Over my head is rushing the stream
that will carry me away.

Oh, to skim the surface, glide,
the crests of breaking waves to ride!

But I cannot, nor stop this tide.
Clinging to the rocks, I hide,
afraid to float, so yet must stay
trapped in circles of night and day
underneath a torrent.

EASTER

Nothing moved. The dull, opaque
ice that stilled the winter lake
stopped me, too; and yet I saw
clearly in that frozen space
where no waves break.

In the leafless London square,
a mist of almond blossom.
Delicate, pink and Japanese,
life returning to the trees;
sex in the air, everywhere!

The churchyard where the crocus blooms,
gold and purple in the gloom
of ancient graves which surely must
beneath, by now, be only dust -
the garden of the empty tomb.

Springtime thaws the frozen stream
and I can ache again.
Rainbows blaze and fade and tease,
the madness of March infects my dreams,
and death is over now, it seems.

WHITSUN

This new life burns me away -
I have disappeared.
It sweeps away the everyday
that I had learned to love,
gives off vapours, steam, a cloud
to carry me above.

Rising over veils of mist
to the white sun clear,
and everything that ever was
or will be, it is here.
But I leave a whisper in the wind
as I disappear.

THE CHANGE

The stormy sea of sex, whose power could lift
to ludicrously happy heights of joy,
obsession and possession, let me drift
with secret smiles on sunlit waters, still
as hazy August afternoons, turn tide
to shatter on the rocks my misplaced trust,
my dreaming hopes, my vanity, my pride,
now casts me on the shore of a new land.

Like Christopher Columbus, I'm surprised;
it's not the place I thought that I would find.
I've never seen such clear and cloudless skies,
such fine-drawn landscapes never met my eyes.
I've crossed the ocean now, and clearly see
an unpromised land, the land of the free.

TROJAN HORSES

You can shower me with praise -
I'm protected from its rays.
Compliments run off my skin,
never find their way within.
They sparkle and are gone, as fast
as raindrops on a pane of glass.

Unless, of course, I love you.

Yet any slight, however small,
penetrates my outer wall.
Sure and sharp the arrow darts,
piercing straight into my heart.
The glass that can no longer flatter
nevertheless will crack or shatter.

Especially if I love you.

Why should I be afraid to believe
the good? Do I think that they deceive?
It seems I've developed immunity
against the dangers of flattery
but not to animosity,
or anyone not being nice to me.

I'm immune to love.

Am I afraid that vanity
may lead to some insanity
of self-delusion? Perhaps it's just
that I'm more likely to distrust
a smiling, potential enemy
than this new strain who does, I see,

make no pretence of love.

THOUGHTS

HIGH TIDE

The rough waves shatter on the land
make little rivers in the sand
then they go.
Maybe we are broken waves
who individually behave
as if our lives are all our own;
froth and bubbles, bits of foam,
each believing "I am me"
as we flow back to the sea
losing our identity
to ebb or flow, eventually,
we don't know.

THE VEIL

What is the thin, strong, misty veil
that masks reality?
One more grasp and I will know,
I will have the answer!
Should I die now, I won't have lived,
for if I cannot see
beyond the swirling chiffon mist,
I am not even born.
I won't stay wrapped in the cloud to dream,
be content with how things seem,
but rend the veil, break through the shell,
be born.

PARADOX

I used to wonder, why are we here?
But when I discovered that life has no purpose,
that purpose was suddenly, dazzlingly, clear.

INNOCENCE

Baby, now you see the world
and all its magic, everything.
Visions I no longer see...take it all in!
For soon you must begin
to walk and talk and read and write,
be civilized,
and all the knowledge you have now
you must forget.
Just for now, you see it all;
keep it safe within,
and in the crowded years ahead,
I hope that in your life will be
of this, a flash of memory.

DESIGNER LABELS

A ribbon of silk, the river,
pale grey with gold-dappled light.
they will see jewels a-glitter
from the plane flying over tonight.

No length of silk has luminous gleam
like water's gold or pearly sheen.
There are no jewels that sparkle bright
as London from the air at night.

But if I said "your dress is like
a strip of muddy water. Such
a necklace! Is it street lamp lights?"
you wouldn't like it much.

PHOTOSYNTHESIS

Why should we prize a shining thing,
a sapphire or a diamond ring?
Do we believe we capture light
to keep against the barren night?

As we turn towards sunrise,
on light we feast our hungry eyes
as flowers photosynthesise,
make nectar sweet, inside.

CROSSED WIRES

Lighten the alleys, the subways of the city of the mind,
that thoughts, born in bright-coloured rooms inside
may not be damaged, injured on their way
through twisting maze of streets to light of day,
the drawbridge of the fortress of the mind -
perhaps to meet another of a kind
in the outside world, where roads are wide and straight
and harsher lights than these outshine the day,
light up the night, yet cannot penetrate
the walled medieval prisons where we stay,
each locked within our little cells, it seems,
with secret fears and darkness and trapped dreams.

TOFFEE APPLE

We have tasted Eden's fruit and know
that we must die; and yet still vainly try
to wash away that apple's bitter taste.
Building empires, seeking fame and fortune,
whatever seems a reason to be here,
and as the glossy golden toffee crumbles
that sugared once its sharpness with a faith
assuring us that we would live forever,
in scientific theory, philosophy, we search
hoping perhaps to find a sweeter fruit.
Or rush around, afraid of looking down
at fallen apples rotting on the ground,
for we are frightened animals who see
the waiting earth beneath the apple tree.

DO NOT DISTURB

Terrified, go hide inside,
don't let them take you for a ride,
from home sweet home you must not stray
but stay with old familiar ways
in any weather.

Safe within your plastic bubble,
insulated from all trouble,
filter out unwanted light
with lacy curtains patterned white
with feathers.

Stick rigidly with your own kind
no virus will invade your mind
with new ideas which might take hold,
break down the iron bars that hold
you up together.

What crime did you commit, to be
so locked away? Could you be free?
Or is your cage a tortoiseshell,
a refuge, not a prison cell,
to be together?

If anyone should ever dare
to pierce with bright unfiltered glare
your curtains, try to change your cage,
you'll snarl and bite, draw blood and rage
to stay tethered.

LITTLE PIGS

I built my house of bricks,
it is my castle, here within
everything's in place.
Tea-bags here, and cornflakes there,
neat and clean and tidy.
I have a proper well paid job,
my holiday is all booked up.
I have it sorted.

They built their house of straw,
and went to live with dreams,
those people on the TV screen
and in the Daily Mail.
They're all made of video-tape
and paper,. they're not flesh and blood
like me, but only fairy-tales -
they aren't real.

They built their house of cardboard,
those people, sleeping rough.
"Any spare change, sir?" Sorry, no-
you chose to live this way.
All these people on the dole,
why should I support them?
They don't want to work, like me.
It's not my problem.

A burglar-alarm's my sentry
keeping watch at night;
my TV screen's a narrow slit
in my thick castle walls.
Everything is bottled, clearly
labelled, put away.
I won't let in the Big Bad Wolf
should he call this way.

LOW TIDE

I stand on the sea-bed among the rocks
their black plump-beaded, seaweed locks
for the creatures who wait for the incoming tide,.
somewhere to scuttle away to hide
from me, the stranger, in a world
twice-daily by the tide revealed.

Primaeval, underneath the sea,
and yet it is not strange to me.
It stirs an ancient memory
I can't locate. I used to be
part of this, aeons ago,
drifting, floating, tidal flow.

And many a chance mutation since,
by tiny errors in reprints,
here I happen now to be,
beside the shore, a castle see -
by chance mutation, adaptation -
upstream, the nuclear power station.

BIMBO KNOWS

It's said that language, speech, evolved
to enable us to lie;
deception flourishes in the gene-pool,
cheat or die.
Chatter, chatter, pitter, patter,
words - they do not really matter.

It doesn't matter what you say,
the cat knows this;
he knows you pass the time of day,
or if you're purring,
he hears what you really feel,
understands what's true and real.

We listen to the surface ripples,
hardly hear the sea.
They hear deeper, powerful currents
better than we.
Talk to Bimbo, he will show
what we've forgotten, that he knows.

PLACES

SHORE PATH WALK

Clear as the spring at Walton Bay
the Welsh hills on a rainwashed day -
mountains glimpsed through the dip in the ridge,
spiderweb span of the Severn Bridge -
distant Exmoor, transparent, faint,
a watercolour wash of pale blue paint.
All so clear, till the sun breaks through,
solid land dissolves and vanishes from view.
The path's a strung-out woodland glade
where hills emerge like dreams, and fade.
A city in a spotlight, then
appears and disappears again -
rippling glittering, golden light
on the sea a dazzle bright
like love.

Mirror-water, getting high,
reflections in a lover's eye -
that is why it matters so,
just because it flatters so.
Narcissus blooms beside a lake,
but ships go by, and in their wake
the glass shatters. Over the bay
light, red-shifting, moves away;
Denny Island stretches wide
reaching for the ebbing tide,
Newport shimmers, a necklace of light
lending the surface of the night
an urban glamour. And yet, listen;
as the silver water glistens
something whispers.

THE GRAND CANYON

Patterns of shadows of clouds were cast
adrift across the silence vast.
Strata, mathematical
horizontal stripes revealed
layers of time in land,
the ages shrank to a moment, and
I was a grain of the desert sand.
A speck of dust which will expand,
sparkling, sometimes, in the sun
of many eyes, or one...
But that was just a moment, too,
the lightning-flash wherein I knew.
When the mist cleared
and every shadow disappeared,
every age was present here
across vast spaces, reaching near
enough to touch, a Promised Land
held within a grain of sand.

THE BLACK MADONNA

As we drove from Barcelona
to the mountain monastery
of Montserrat, we clearly saw
why poppies remember the dead of war;
the fields were splashed with red.

But we were here to enjoy the view
of the jagged mountain, buy postcards,
see whatever there was to see,
have a glass of wine, maybe
among the other tourists.

Candles blazed in the gloom of the church
heavily fragrant with incense, when
we somehow found ourselves in line
to kneel before a special shrine;
the shrine of the Black Madonna.

Something evil emanated
from that blackened, ancient doll.
Malevolent, Satanic power,
hatred on a bank of flowers;
if there is a Devil, I thought, he is here

in a church! But history here
is soaked in poppy-scarlet blood.
This wooden doll, this graven image,
in a cave for centuries
hidden from the Moors.

For Christianity, a focus.
Symbol of the soul of Spain.
Fifteen centuries of pain
so absorbed into her frame
that now, she spoke for herself.

Radiating down the ages
cruel, obscene intolerance.
To such a thing, how could they pray?
We were glad to get away
to the poppy fields.

ALFRED PLACE

Under the dome of the wide Suffolk sky
the spirit spreads and opens out
like blossoms in the orchard;
apple-tree brides in cowparsley lace
at their communal wedding.

The pond in summer, where dragonflies hover,
the willow weeps, the waterlily smiles.
Emma's pony in his field
reaches for the greener grass
beyond his fence.

Safe and comforting, the house
wraps around us like a blanket.
Scent of pine-logs, chestnuts roasting
on the fire. The inglenook.
Cosy as cocoa.

Winter sunrise draws an outline
neon red on snowy fields.
The kids make a snowman on a white lawn
as smooth as the icing on the Christmas cake,
but for their footprints.

Oscar the Irish wolfhound joins us
New Year's Eve for Auld Lang Syne.
Lopsidedly, the children's snowman
disintegrates and melts away
with their footprints.

THE SIRENS

Between the rugged pine-drenched coastline
and cloud-streaked blue Vancouver Island,
along the Straits of Juan de Fuca,
in our little roller-coaster
tossing, jumping, over waves
past Cape Flattery we chugged
to the calmer Pacific Ocean, where
we hoped to catch a salmon.

A mile or two off Tatoosh Island
we switched off the engine of our boat.
To the west, no land until Japan.
Golden sky curved around us
to meet the reflecting gold of the water
swelling beneath us, and we heard
the strangely soothing music
of the wild tranquil ocean

rising, falling, misty rhythms,
singing voices of sea and wind...
Distance was here and far away
and it seemed presumptuous of us to speak
while such sea-angel voices sang
of all that ever was, will be,
echoing eternity
in hypnotizing harmony.

The mist came down quite suddenly.
We could see no land at all.
Starting the engine we broke the spell,
no longer heard the Sirens wail.
We caught a cod for dinner next night,
returned to the harbour with yellow lights
and bobbing boats and noisy folks,
laughter and fisherman's stories and jokes.

THE EXILE

Between the steep and wooded cliffs
the rushing river Frome
tumbling over waterfalls,
glides around the island, Bamboo Island,
under the one-man bridge,
over the weir by the watermill,
sepia-tint through Snuffmills.
Beyond the trees at the top of the hill
is the house where I grew up;
someone else's, now.

In town I explore the Bristol streets
where Frome and Avon meet,
under fountains set in concrete.
Corn Street, Park Street, College Green,
King Street, the Llandoger Trow.
People meeting friends for lunch,
rushing back to work,
waiting to catch the bus to go home
to the suburbs, as I once did;
I'm on the outside now.

I look around the Arnolfini,
St Mary Redcliffe Church,
a tourist in my own home town.
And everywhere it seems they say
you don't belong here now!
You did not stay! I hear them say,
You left, you went away.
This city is no longer yours;
you chose to close the door.
You've been gone too long.

THE SHOW MUST GO ON

In the pearly glow of the African dawn, I awoke
under mosquito-net veils, in my tree house.
The beating of the drums announced sausages, bacon, cooking
on an open fire. We set off in the jeep
into the bush.

With the rising of the sun, the humming of the insects rose,
in tune with the intensifying colours.
Rounded sounds of warbling birds - an opera!
Curtain up. No parts for us in this show.
We're the audience.

All day we gazed in wonder at impala,
zebra, baboons, a crocodile. Elephants
silently swishing water over their bodies.
The gathering at sunset at the water
to drink together.

And they are all in tune with their surroundings,
synchronized, in perfect harmony.
Elephants create the waterholes,
warthogs plough the soil before the rains,
and seeds grow.

An elegant giraffe once looked towards us
imperiously, incurious, as royalty
might glance upon a crowd of ragged peasants
or the paparazzi; cameras clicked
and whirred and snapped.

We are humbled here, somehow, outsiders.
This is their world. An exclusive club
To which we never, ever can belong.
This is the garden which we long ago
were cast from.

We cannot return; we've come too far
along the road of tinkering with nature,
with agriculture, industry and science -
but we must let them stay the way they are,
in the garden.

GREAT COURT AT THE BRITISH MUSEUM

It is a splendid, modern space,
daylight, blue and curving,
gathering galleries apace
around the Reading Room,
whose dome of lilac, cream and gold
soars over where great minds have read,
still centre in this house of old
treasures of antiquity,
surviving of humanity.

The curving, latticed roof of glass,
a space-time diagram, which draws
together present, future, past,
pulling in the sky,
flooding light and harmony
into an inner pool of light
with colours of the sea -
blue, sea-green or grey
varying each day.

Why does it move me near to tears?
Because this century of ours
now gives a gift from our own years
to all those other ages.
We make to Egypt, Greece and Rome
a tribute; and now they look on
before we too are gone.
And I am struck with wonder how
it is my luck to see it now.

CLOUDS

Away the station platform slides,
smoothly past the houses glide,
and so too my troubles hide
and seem to slip away
in green and yellow fields.
Woodland, downland, steady riding,
sudden flash of railway siding.
Undulating, not quite still
blue horizon, gently gliding,
takes the slower rhythmn in this dance of hills.
I drink cappuccino and doze and dream,
while overhead they sail serene,
the billowing, following clouds.
The clouds are coming with me.

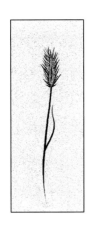

PART THREE: AND THEN..

BURIED TREASURE

RED SHIFT

BURIED TREASURE

PREMONITION

When I was little
I thought the red pillar box
knew by some magic
where to send the letters.
That the postbox could read the address
and whizz it there by magic,
travelling underground
along mysterious passages
all the way to Uncle Dennis
in Oxfordshire.
That there must be an underground network
of tunnels to every house,
spreading out like spider's webs
from every postbox.

Then I couldn't understand
why we needed the postman.

What silly ideas I used to have,
so long before E-mail.

BURIED TREASURE

Remember spangled pavements?
Diamonds used to sparkle there
inlaid within each concrete square.
I saw them when I was little
before I knew
a pavement was just a pavement.

I look at the pavement now;
across the street the lamplight spills,
there my diamonds glitter still
but I'd stopped seeing them
because I knew
a pavement was just a pavement.

THE MAZAWATTEE TIN

The photographs are crammed within
an old blue Mazawattee tin;
sepia tint, Victorian faces
look at me through time, from places,
moments frozen long ago
in the photographer's studio.

Caroline and Clara, James,
Charles and Sarah, only names,
strangers, destined yet to share
great-grandchildren, gathered there
down at the bungalow at the Bay
on a 1930's summer day.

Together for a moment, trapped
in tiny, tattered, black-and-white snaps.
Released, they drift apart, won't stay,
out with the tide they are carried away;
a family spreading out to sea,
strangers once again to be.

But in the Mazawattee tin,
there the tide is always in.

THE LOST BIRTHDAY

They said I would sleep. It wasn't sleep, there was no
gentle breaking up of floating thoughts,
kaleidoscopic images, sliding
into dreams, no drifting up and down
the levels of sleep.

At 9am they put the mask on my face,
told me to smell some perfume, and held me down.
I struggled against the scent of lavender-water,
the buzzing in my head - a television screen
full of interference

(the hissing of the cosmic background waves,
from the dawn of time, from the beginning).
Instantly it was four in the afternoon.
I was back on the ward, a pain in my throat, thirsty,
and minus my tonsils.

No time existed in between, they'd taken
my tonsils out, somewhere outside of time,
for I'd jumped straight to 4 o'clock, and missed
that day - which happened to be my ninth birthday -
I'd taken a time-trip.

I wonder if this is how death will be, at the last,
a losing of consciousness into the blur of microwave
radiation? And if, should there be
another life, a million years ahead,
I'd jump straight there, too.

THE SMILE

I ran across the playground to my mother
waiting for me at the school gate
with a lady with a pram, whom I didn't know.
"Come and meet Janet!" said Mummy to me.
Janet was a bright-eyed baby,
sitting up in the pram.
I gave her my finger to hold.
How tightly her tiny fingers grasped!
Her face lit up with a shining smile,
so delightedly,
as if she recognized me
as being different from a grown-up.
Nearer, perhaps to her.
But I was as old and as far away
as they were.
"She likes you!" Janet's mother said.

That summer I played in the garden one day
my mother came out of the house.
"Wendy.." she said, and I saw that she cried,
"You remember Janet, the baby you liked?"
and told me that Janet had died
and what the letter had said.
"The doctors did everything that they could"
the baby's mother had written
"but her little body could not stand
the operation."
That memory stings my eyes
fifty years on. I can still see
Janet's happy smile
when she recognized me as what she might be
but never was.
The smile she gave to me.

THE BUNGALOW

No mod cons at all, no plumbing,
no electricity.
It stood in a corner of a field
overlooking the sea.
(The Bristol Channel, actually
the Severn Estuary.)

Apple trees in the garden, and nettles
right up to my window-ledge.
Thrushes sang in the mornings
where brambles entwined in the hedge
Rambler roses climbed to the edge
of the wooden verandah.

The verandah! Where we sat to watch
the big ships passing through
the gold path painted by the sun
late in the afternoon,
and there was no-one that we knew
who had a bungalow.

The smell of slowly roasting lamb
and garden mint on Sundays.
Under the table, the buckets of cold
spring water from the Bay
along the shore-path a few minutes away.
The rainwater butt.

A smoky fire, some oil-lamps,
no still electric light,
but candles throwing shadows,
liquid, living light
moving, flickering, alive,
enveloping ghosts.

I have another place there now
with water hot and cold
and all those things we think we need.
Ships sail through the gold
upon the water as of old,
but something is lost.

SURVIVAL

The mist that cleared, just once for me,
swirls around, and I can't see.

I cling to rails of what has been,
a skeleton, no flesh between
the bones of ritual and routine,

Four o'clock is time for tea.
Decorate the Christmas tree.
Pretend to be alive, until
against my will, for good or ill,
life grows back.

Because it will.

FAMILY HOLIDAY

Food fresh-plucked from local soil,
tomatoes, garlic, olive oil,
the market in the village square,
alive and sharp and real. Compare
with dumbed-down, shrink-wrapped plastic halls
of antiseptic shopping malls
where senses blur and merge and slide,
tranquillize, anaesthetize.

Which is real, the holiday?
Or bland, familiar everyday?
Steep cobbled villages, floating Pyrenees,
cafe in the shade of the chestnut trees.
Table set for lunch by the pool.
lapping, rippling, blue and cool,
stretched out bodies, stretched out minds,
music of cicadas, taste of wine.

Like any proper family
we do not always quite agree.
Why you all do the Lottery
remains a mystery to me -
I think it's a cynical government plot -
but I see little harm in GM crops.
And truly, now I'm that much older,
sex and chocolate leave me colder.

Superficial changes, these,
like aches and pains in backs and knees.
We're getting older all together,
been through much more changing weather.
We're a family of friends
with whom we never need pretend.
But it's gone, like a brief intense romance -
our summer in the south of France.

THE CRUELLEST MONTH

Forsythia bright
filling my room with yellow light;
pure delight
in an old man's face
at the sight
of cherry blossom,
palest pink and fragile, dreamy,
exquisite, yet rich and creamy,
visual ice cream.

Under the tree
a toddler romping joyously
amidst confetti
in the grass;
serendipity
in a pink snowstorm.
Already the petals fall like snow
as autumn leaves fall, south, below
the curve of the earth.

Sharing joy,
the old man and the little boy.
Sunshine. Enjoy!
By the boating lake
I have coffee and cake
in soft April air.
Whispers of green on gnarled old trees
and saplings swaying in the breeze.
Sudden rain.

CHRISTMAS TIDE

The tyranny of Christmas, how
I wish I had the nerve
to just forget the whole damn thing,
hibernate until the spring,
not send a single card.

Why should I be expected to
spend money I don't have
on useless things for others who
will give me things I cannot use
and do not like?

The crowds of Oxford Street, the shops,
they give me claustrophobia.
I hate all people, young or old,
and anyway, I've got a cold,
and they're all in my way.

Then comes the first card in the post;
how nice to hear from her!
Then family and friends extend
an invitation, please, to spend
the holiday with them.

In a country church packed full
we have a jolly sing-song
of carols in the candle light,
then party on into the night
drinking mulled red wine.

In the morning, opening presents
(bought in Oxford Street, before)
the tree scents all the rooms with pine;
there's turkey for dinner, with stuffing and wine,
pudding and brandy butter.

By the sleepy log-fire crackling,
drifting, dreaming, hibernating,
we forget the whole damn thing,
while on the lawn, a hint of spring;
snowdrops coming up.

COUNTRY PURSUITS

Soft in the hollows of the curving countryside, the woods,
shadowy, like downy hair, give cover
to slopes and streams and damp dead leaves and mud,
secret pathways beg to be explored,
entered, torn through, ripped; asking for it.
Against the winter sky the lacy edge
is broken, frayed. Blood runs in the stream.

In red coats on groomed and gleaming horses
longing for the chase, the cheerful riders;
the hounds as one obey the sound of the horn
and take their place; clip-clop the others follow.
The way to spend a winter day! But turn
your mind away, and do not stay to dwell
upon the fox you send to pre-death hell.

RED SHIFT

INVISIBLE WOMAN

Once, I could have made your day
with a smile.
You made fools of yourselves
just for a glance
from me.

And if I agreed to spend an evening with you,
any of you,
you'd fluff out your feathers
and lose no chance
to boast to me

how clever and brilliant and famous you were,
how attractive,
so I'd realize how lucky I was
to dance this dance
with you.

And how I despised you, so easily fooled
by eyeliner,
lip-gloss, and fresh blow-dried hair,
you thought me entranced,
gullible.

Were we only showing off?
To whom?
A world careless as all of you
who now in ignorance
look through me,

leaving me to browse in bookshops,
wander undisturbed,
now that I'm invisible,
even when enhanced
with eyeliner?

No. We knew what was true.
But floating through a future
that suddenly arrived,
I am the ghost of the girl who danced
quite recently.

HIS NAME WAS SANDY

My new friend the ginger cat is dead;
I didn't know him long.
He was young and strong;
it was only just last week
we were playing hide and seek
in my garden. Now he's gone.

I thought because I'd lost those I loved most
that I was now immune from pain,
never could be sad again.
But I find that I was wrong.
And so now I grieve
and cannot quite believe
that my new friend, the ginger cat, is dead.

FALLEN LEAVES

Shrouded in fog, the twigs and thorns
like antlers, or like devil's horns
skeletons of trees.

Burning on the funeral pyre
of an autumn garden fire,
fallen dead leaves.

As longer draw the hours of night
let us appease the gods of light
with firework displays.

With glitter-fountains flood the night,
impregnate with shimmering light,
another son beget.

Witches on broomsticks, Halloween masks,
scarlet poppies, the Cenotaph -
"Trick or Treat?" we pray.

Yet still more the daylight fades;
in the mists are ghosts and shades
whispering "Trick!" they say.

And every dusk November, when
we silently remember them,
they silently forget.

THE SOMME REPLIES

I was alive once, too. Briefly.
My two eyes filled with moving pictures bright
of faces smiling welcome, drawing me
in across the threshold into light
from some dark deep forgotten sleep, some night
where I was before. Why? What for?
Your promises you didn't keep. You lied.
Because of you, your love, your hate, I died.
Horribly. Don't kid yourself, there's nothing
glorious or noble in my suffering.
And now you say you're sorry, now I'm gone,
that at the going down of your gold sun
and in the morning, you'll remember me
briefly, in a solemn ceremony.
Well, good for you.

DADDY

You let me stay up, just five more minutes.
Spoiled me, they said.
Playing bus-drivers in a chair,
telling me stories,
chocolates on Fridays, if I was good
(I always was).

Down at the bungalow at the Bay
on our holiday,
drinking water from the spring
on the beach,
counting every seventh wave
at high tide.

Taking our dog Curls for walks,
long wet grass,
waving to ships sailing close by
from the verandah.
Watching the sunset over the Channel,
the smell of oil lamps.

Christmas time and lots of parcels
from Father Christmas.
Aunts and uncles playing games
at the party.
We children staying up late,
drinking raisin wine.

In my fiery adolescence,
did we quarrel!
Wonderful, dramatic rows
like in movies.
But we always made it up,
didn't we?

Drama classes, then, for me,
dreaming of Hollywood
but singing songs around the piano.
You were the star!
Everyone said you sang as well
as Bing Crosby.

Meeting me from the London train
at weekends,
you'd tell me all the family gossip
in the car,
then pour me out a gin and tonic
when we got home.

Suddenly you were very sick,
I could see.
But fatal illness only happens
to others.
Other people's fathers die -
not mine.

I was lonely when you died,
panicky,
never felt so frightened before.
But that's the price,
and if this pain should never ease,
it's still a bargain.

KATH

People were always dropping in
for Kath was cheerful and kind.
Welcome shone in her hazel eyes -
they knew she wouldn't mind.
But she was never quite so cosy
as her rounded, plump and rosy
image implied.

A clever mimic, she used to make
the family fall about
with spot-on impressions of visitors
as soon as they'd gone out.
so we forever more could see
the ridiculous side of humanity -
including ourselves!

As a girl she had hoped to study art;
no future in that, they had said.
She'd got a job at the pottery
painting cups and saucers instead
until she got married. Not so much a waste
as diffusion of talent into good taste
in home and garden.

Her kitchen was full of good things to eat;
there was always a home-made cake.
Around the house were works of art
that she had learned to make
at her evening pottery classes,
and garden flowers arranged in vases
beautifully.

Kath loved animals and flowers,
and English countryside,
and me. She was my greatest friend
in whom I could confide
in laughter or in dark despair.
I always knew that she was there.
She was my mother.

GOOD FRIDAY

The bungalow's black skeleton,
cremated, naked, starkly stares
out across the white-frilled waves
that break, re-form, and break again.

Long buried under strangers' changes,
of flesh of home-improvements bare,
the charcoal frame, now stripped and spare,
reveals the ghost of what was there:

a summer home with apple trees,
long grass, and meadow-flowers. There
red hens clucked and gossiped, where
now lie like tombs the concrete squares.

Echoes of a family
long-gone, no longer anywhere,
whose everywhere-and-nowhere waves
break, re-form, and break again.

One who neither knew nor cared
was paid to set this place alight.
Thirty pieces of silver bright
to burn it down, the other night,

to make way for new park-homes there
with panoramic Channel views.
Bungalows re-formed anew.
One day someone will love them too.

Always look on the bright side.

EPILOGUE

Closing my eyes I drift away
to a land with pink sea,
where cats can talk and I can fly,
floating in air, set free.
Dreams, unlocking the doors of their cells
dark in my mind and deep
dance through the night, and when I awake,
the dreaming dreams will sleep.

A WORD ABOUT WORDS

A layman's glossary of scientific terminology which has been used in these verses.

1. Multiverse

In Everett's "Many Worlds" interpretation of quantum physics, the universe is one branch of many alternative universes, existing at right-angles to each other, which split and branch off from each other every time a universe is faced with a quantum choice. In this theory, all probable worlds actually exist, but are cut off from each other. This multiplicity of world or universes is often called the multiverse.

2. Local realistic view

The common-sense view of the world, whereby everything has objective reality, and exists regardless of observation, where time flows from past to future, and where no influence can exceed the speed of light. See John Gribbin's "In Search of Schrödinger's Cat", pp 222-224. Unless otherwise stated, subsequent references to Gribbin are to this volume.

3. Electron spin

The spin of an electron was measured to test Bell's Inequality (see following) in a real experiment.

4. Bell's Inequality

The Bell Test, named after John Bell, starts from the assumption that the local realistic view of the world is the true view, and calculates that if this is true, the measurements of electron spin should result in a particular inequality. If this does not happen, then the local realistic view is false. All tests as of today have shown that this is indeed false, and the quantum mechanical view of the world is more probably true. See Gribbin, above, chapter 10.

5. Wave-particle duality

Electrons and photons behave as waves when not being observed, and as particles when observed or measured, when the wave function collapses. See Gribbin, Chapter 5.

6. Erwin Schrödinger's Cat

Erwin Schrödinger's famous imaginary experiment in which a cat is enclosed in a box with a phial of poison, wired to a device with a 50/50 chance that the poison will be released so that the cat dies, or not, and the cat lives. The Copenhagen Interpretation of quantum mechanics states that until someone looks in the box, the cat is neither alive nor dead, but somehow suspended in a juxtaposition of states. Only when someone looks and collapses the wave function to find either a dead cat or a live cat does the cat's fate have any reality. In Everett's "Many Worlds" interpretation, however, the universe splits in two, and in one universe there is a dead cat, and in the other a live cat, regardless of observation, and both cats are equally real. Applied to every probability in the universe this implies that the universe is constantly splitting into an infinite number of universes. See Gribbin.

7. Uncertainty Principle

Heisenberg's uncertainty principle states that it is impossible to know both the position and momentum of a particle at the same time, not only because the act of measurement would change where the particle is or where it is going, but because, according to quantum mechanics, a particle cannot have a precise position and a precise momentum simultaneously.

8. Cosmic background microwaves

This is the microwave radiation left over from the Big Bang, which is distributed evenly throughout the universe. We see and hear this echo of the birth of the universe on an analogue TV screen when it is not tuned in properly. The screen is full of flashing dots and the sound hisses, as does a radio between stations. Instead of picking up a TV programme, it is picking up and amplifying this cosmic background microwave radiation.

9. Photon double-slit experiment

An experiment in which photons are passed through a screen with two holes or slits onto a detector screen. When both slits are open, an interference pattern of dark and light stripes appears on the screen, showing that the photons are behaving as waves, overlapping and interfering with each other just as waves in water do. When one slit is closed, the photons behave like a stream of discrete particles, or billiard balls, and the screen shows a pattern consistent with this, and with no interference.

This holds true even when only one photon passes through the experiment at a time. When both holes are open, even one photon shows the interference pattern, as if it somehow passed through both holes at once and overlapped with itself. See Gribbin, Chapter 8.

10. Transaction theory

This is John Cramer's transactional interpretation of quantum mechanics which allows faster-than-light travel for photons - and a photon travelling faster than light is travelling backwards in time. Cramer postulates that some of the overlapping, interfering waves in the double-slit experiment are interacting with waves from other parts of space-time, including the future, to send messages back from the future to the present, through action at a distance. See Gribbin, epilogue, "Schrödinger's Kittens".

11. Quark

The smallest known constituent of matter, on present knowledge.

12.. Brane

Short for membrane; in a version of Everett's "Many Worlds" theory, the multiverse consists of "Many-folds", where each universe is a membrane lying side by side, and folded over the next, but invisible to each other. See "The Edge of Infinity", Alison Boyle, New Scientist, September 2001.

13. Red-shifting light

The Doppler shift. When light shifts to the red end of the spectrum, and the wave-length is longer, the source of light is moving away from the observer.

INDEX OF TITLES

Alfred Place, 61
Amber Light, 34

Beyond Our Brane, 26
Bimbo Knows, 57
Black Madonna, The, 60
Black Widow, 44
Bungalow, The, 74
Buried Treasure, 70

Change, The, 48
Christmas Tide, 78
Clouds, 67
Control Freak, 43
Country Pursuits, 79
Crossed Wires, 53
Cruellest Month, The, 77

Daddy, 84
Daybreak, 12
Depression, 46
Designer Labels, 52
Diamonds in the Sky, 27
Divided we Stay, 31
Do Not Disturb, 54

Easter, 47
Ego Trip Through a Black Hole, 28
Epilogue, 88
Escape, 46
Everywhere At Once, 20
Exile, The, 63

Fallen Leaves, 82
Family Holiday, 76
First Light, 33
Footprints in the Snow, 21

Good Friday, 88

Grand Canyon, The, 59
Great Court at the British Museum. 66
Green Light, 41

High Tide, 50
His Name was Sandy, 81

Innocence, 51
Invisible Woman, 80

Just Looking, 35
Just Sex, 39

Kath, 86

Leap Before You Look, 18
Little Pigs, 55
Lost Birthday, The, 72
Low Tide, 56

Magic Circles, 23
Magic Gene, The, 30
Many Worlds, 16
Mazawattee Tin, The, 71
Multiverse, 11

Nothing, 29

Out of Order, 36

Palpable Hit, 35
Panthropic Principle, 25
Paradox, 51
Photosynthesis, 52
Premonition, 69

Quanta, 13

Red Light, 37
RicePaper Moon, 40

Searchlight, 24
Separation, 45
Shore Path Walk, 58
Show Must Go On, The, 64
Sensible Thing, The, 42
Silver Scribbles, 3
Sirens, The, 62
Smile, The, 73
Somme Replies, The, 83
Stay Tuned, 22
Survival, 75

Toffee Apple, 53
Transfusion, 38
Trojan Horses, 49

Veil, The, 50
Virtual Reality, 17

White Horses, 31
Whitsun, 48